Copyright 2020 Gumdrop Press

All rights reserved.

No part of this book may be reproduced in any written, electronic, or photocopied form without written permission of the publisher or author.

Every effort has been made to ensure the accuracy of the information contained in this book. The author and publisher disclaim liability to any party for any loss, damage, or disruption caused by errors or omissions that may result from the use of information contained within, whether such errors or omissions result from negligence, accident, or any other cause.

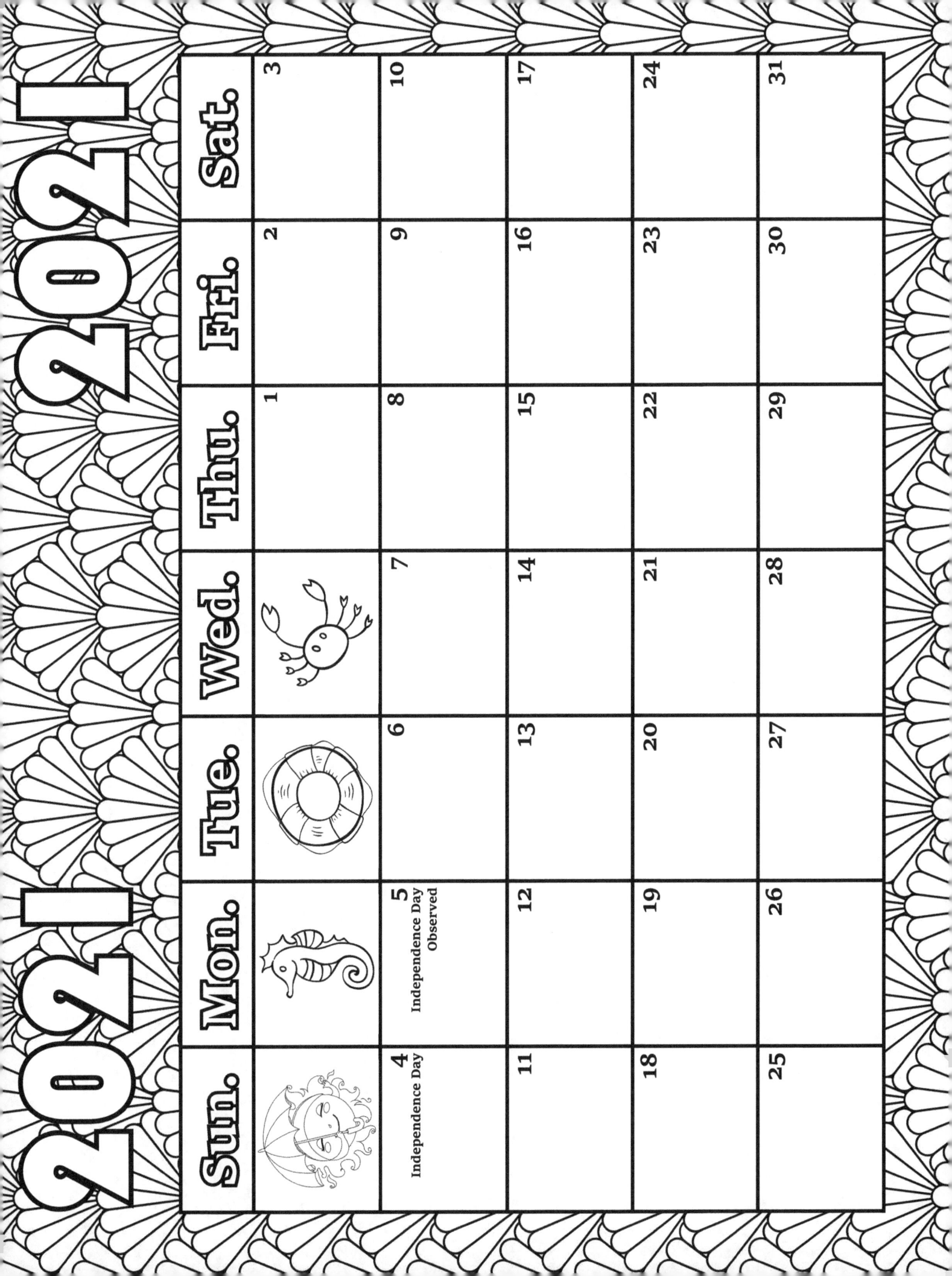

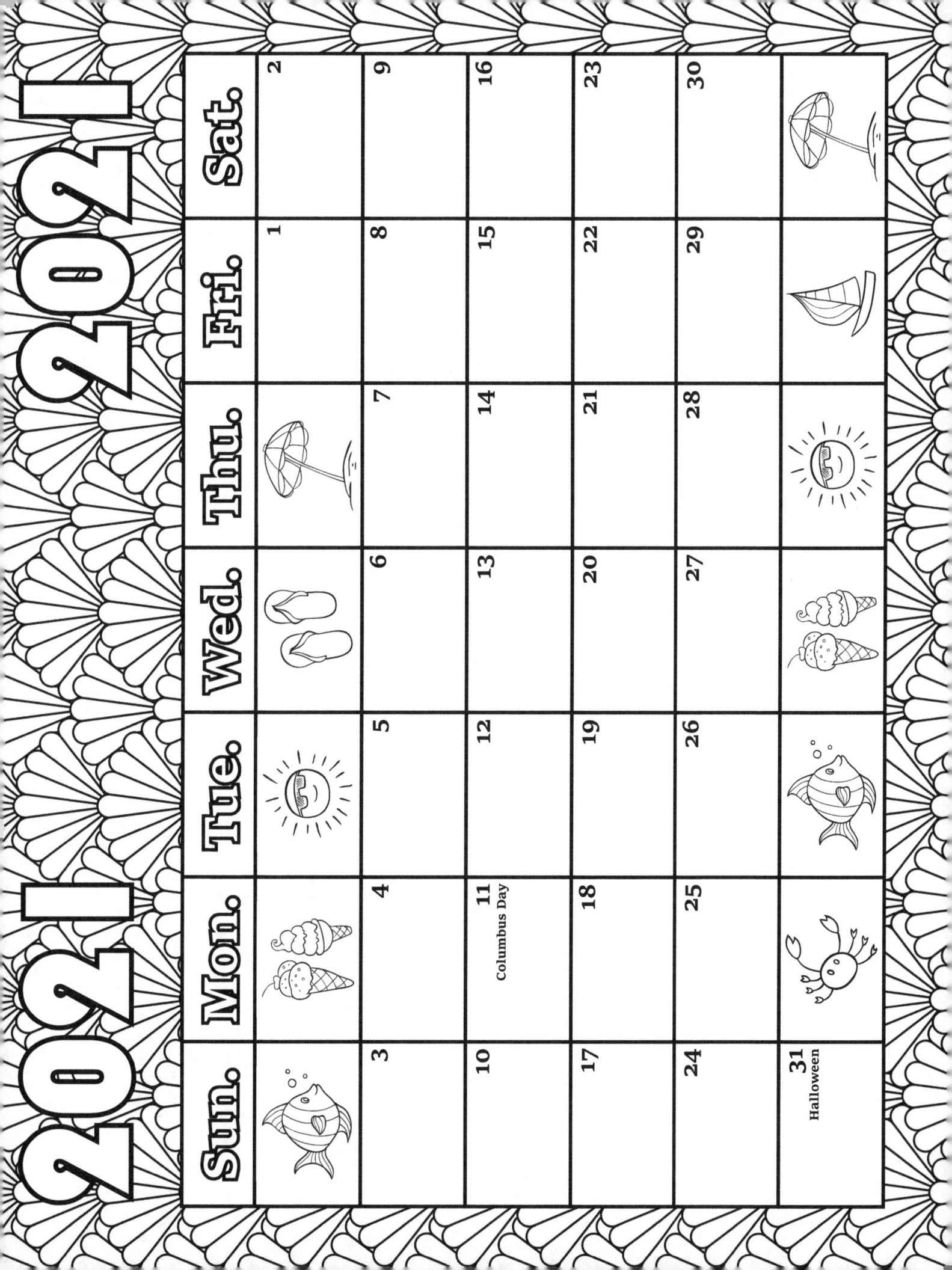

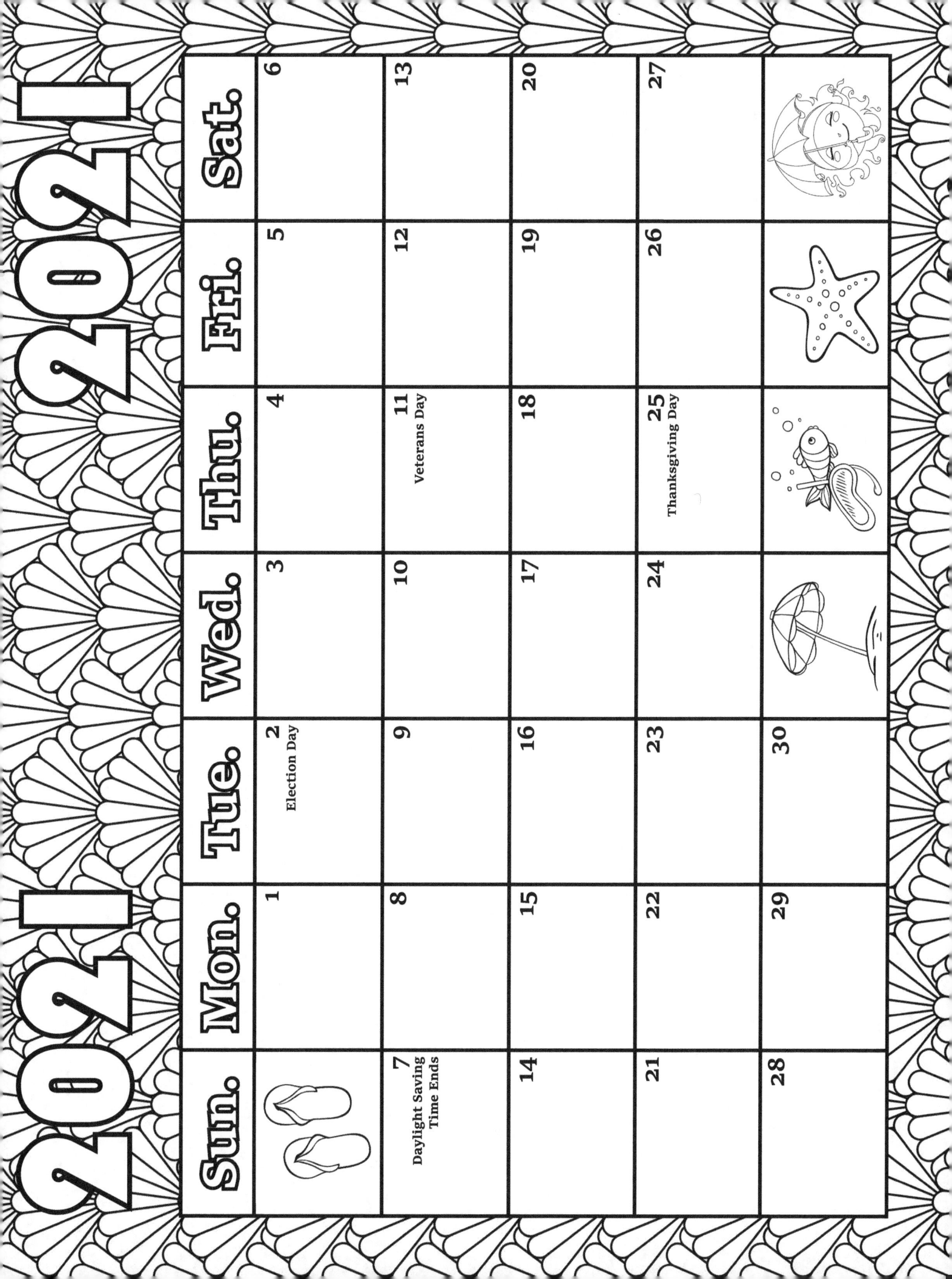

www.ingramcontent.com/pod-product-compliance
Lightning Source LLC
Chambersburg PA
CBHW081128080526
44587CB00021B/3795